Emile OKONGO LOKOLONGA

WORKING CONDITIONS AND PATIENT SATISFACTION IN SANKURU

Emile OKONGO LOKOLONGA

WORKING CONDITIONS AND PATIENT SATISFACTION IN SANKURU

Imprint

Any brand names and product names mentioned in this book are subject to trademark, brand or patent protection and are trademarks or registered trademarks of their respective holders. The use of brand names, product names, common names, trade names, product descriptions etc. even without a particular marking in this work is in no way to be construed to mean that such names may be regarded as unrestricted in respect of trademark and brand protection legislation and could thus be used by anyone.

Cover image: www.ingimage.com

This book is a translation from the original published under ISBN 978-620-3-41838-5.

Publisher:
Sciencia Scripts
is a trademark of
Dodo Books Indian Ocean Ltd. and OmniScriptum S.R.L publishing group

120 High Road, East Finchley, London, N2 9ED, United Kingdom
Str. Armeneasca 28/1, office 1, Chisinau MD-2012, Republic of Moldova, Europe
Managing Directors: Ieva Konstantinova, Victoria Ursu
info@omniscriptum.com

Printed at: see last page
ISBN: 978-620-3-48010-8

DEDICATION

Only the diet cutter can make palm nuts available to you. (Sankurois
proverb).

ACKNOWLEDGEMENTS

First of all, I give thanks to Almighty God, Master of time and circumstances, for my courage, my willingness and my ability to keep my balance until I succeed.

Secondly, I would like to thank my supervisor, Bilal Selman, for having supervised me throughout the writing of this dissertation. His comments and guidance were well-founded and beneficial to my improvement.

I would also like to thank all teachers Unicaf University who have contributed this high-quality course, which has propelled me into the ranks of high-ranking sceptics.

Finally, I would like to thank my family who supported me during this long period of study, my wife Bibiche and all my children, for having endured all the hardships caused by these studies.

SUMMARY

The issue of working conditions and patient satisfaction is becoming a priority for the health sector, which is combining its efforts improve the quality of its services with a view to satisfying patients. Talking about quality of care implies patient satisfaction; as long as the patient remains dissatisfied, we will not be talking about quality. Consequently, efforts must be made to achieve patient and family satisfaction.In the quest to improve performance, in a context dominated by the political hijacking of health administration, and also hampered by budgetary restrictions, building the loyalty of healthcare providers with a view to satisfying patients' needs is not easy, although it must take centre stage. That's why it's important to find and implement a new policy to ensure that providers remain loyal and that patients' expectations are met. (JIHANE SEBAI, 2021; Kroger et al, 2007; Donabedian, 1980).The positivist paradigm and the hypothetico-deductive, inductive and adductive approach have been chosen as the best way of getting to the truth.The philosophical approach (using the hypothetical-experiential category), based on ontological and epistemological hypotheses. While the mixed method with the use of the exploratory sequential model and the explanatory sequential model, were implemented to give meaning to our study. The study had three objectives:

1. Determining the impact of healthcare providers' working conditions on patient satisfaction;

2. - Describe the level of satisfaction of patients who consult health facilities in the Sankuru DPS;

3. Determine the current working conditions of health care providers in the Sankuru DPS;

To gather information that answered the research question, we drew inspiration from numerous studies that dealt with similar subjects. Our study was therefore a secondary study that attempted to find answers about the impact of working conditions on patient satisfaction. Our study had experienced limitations that prevented us from obtaining data to analyse and evaluate the last two objectives, as achieving them requires primary data collection. To meet the university's requirements, we used the checklist to extract information on our variables of interest from the datasets of other publications that dealt with similar subjects. The information gathered was encoded and analysed using statistical methods to determine the impact of healthcare workers' working conditions on patient satisfaction. For this purpose, the statistical model of logistic regression was used to determine the significance level of our test. The test revealed that the relationship between working conditions and patient satisfaction is statistically highly significant at the threshold of $\alpha \leq o, o5\%$, thus social economic characteristics positively influence patient satisfaction, as, the p value being well below 0.05. The odds ratio is 0.2926.

TABLE OF CONTENTS

CHAPTER I

INTRODUCTION

The health system in the Democratic Republic of Congo is obliged to manage a plethora of employees imposed on it by political parties in the country's 26 provinces, including Sankuru. This situation not only makes it difficult to provide care for these employees, but also to assess their working conditions. Here and there, the families of patients and the patients themselves are constantly complaining dissatisfaction with the care they receive.

I.1. Problem Statement

The 22-26 July 2019 report by the World Health Organization (WHO) on technical consultations and country action plans, which met in Addis Ababa, paints a bleak picture of child survival in Africa.According to the report, 5.3 million children under the age of 5 died in 2018, including 2.5 million newborns. The report points out that sub-Saharan Africa, which includes the Democratic Republic of Congo, still accounts for more than half of these deaths. The report shows that 52 countries worldwide are still a long way from meeting the child survival target, as set out in target 3.2 of the third sustainable development goal of the global compact (SDG 2030). This threshold stipulates that the child death rate must not exceed 25 deaths per 1,000 births. The DRC is one of 52 countries where the child survival situation remains precarious.80% of deaths occur at home in the DRC, not only because households lack the money to pay for medical care and because of ignorance, but also because patients do not seek care because they not satisfied. 20% of deaths occur in health facilities

(SRSS 2010). For the In the DRC, diarrhoea, pneumonia, malaria and malnutrition are still the main causes of child deaths. (WHO 2019 p1). While target 3.8 of the 2030 MDGs calls for everyone to benefit from universal health cover, which includes protection against financial risks, in the DRC we are still seeing realities that fall far short of this target. And some children and adults die in the arms of their guardians, having been dragged to the door of health care facilities, without having received the slightest treatment because they have no money to pay for medical care.

"Health has no price, but it does have a cost", they say, and this cost must be borne by the state. Today, a student from the Faculty Economics, living in Sankuru, finds it hard to believe this statement, especially when they have lost a loved one in their arms, at the hospital door, without having received the slightest care, quite simply because the family had no financial means to bear the cost of care.

Because of the weakness of certain official state services, such as civil registry, on the periphery of the country, much information relating to the causes of death escapes the control of the disease surveillance department.

In the same vein, in my first practical work for the research methodology course, I pointed out that, based on our own observations, some patients who consult certain health care structures in the Sankuru DPS are not saved and die in cases where death is avoidable due to carelessness or failure on the part of health care professionals. Unfortunately, these latter causes of death remain unofficially published and are therefore attributed to the 5 main causes mentioned above. Pregnant women run the risk of giving birth alone in the field or in their huts, while others are assisted by their peers who have the courage to save lives despite having no training. According to MUSUNGAYI of the POLITICO.CD website, despite the introduction by the President of the Republic, Félix Antoine TSHISEKEDI, of free maternity and neonatal care, in order to align with universal health coversome health professionals in 320 health centres and 50 general referral hospitals in the capital, Kinshasa, see this measure as they were being forced to swallow a bitter pill. Some health professionals are in conflict with this provision, which aims to guarantee access to medical care for both the poor and the rich. Jean-Paul DIVENGI and Roger KONGO, respectively director of the Kinshasa general reference hospital ex MAMA YEMO and the NGALIEMA clinic, were dismissed from their postsaccused having failed in their responsibility to guarantee care of a pregnant woman during childbirth and of her newborn baby. This sanction was imposed in accordance with the provisions of articles 10 and 26 of Ordinance n°81-067 of 07 May 1981 on administrative regulations relating to discipline, following the occurrence of a maternal death due to negligence, at the hospital gate.

This measure continued until the suspension of operations at the Akram hospital in Limete, Kinshasa, for the same reason. The same source maintains that the start of the application of free childbirth and newborn care has not been easy in Kinshasa hospitals. Several corroborating sources who have been on site have confirmed that access to care is selective in favour of those who have the means, while those who do not have the means benefit from the indifference of care providers. MUSUNGAYI (10 Feb. 2024).

On 27/02/2024, a group of health service providers, the head nurses of the health areas in the Lodja health zone, in the Sankuru DPS, appeared in open court at the public prosecutor's office near the Sankuru high court. steal unsaleable medicines from their health facilities and sell them to street vendors.

From this, it can be inferred that the negligence, lack of commitment, ill-treatment of patients, imbalance between supply of healthcare and the demand for it, and all other practices that deviate from the supply of medical care, observed in healthcare facilities in the Democratic Republic of Congo in general and in the Sankuru provincial health division in particular, are the result of the working conditions to which employees in the public healthcare sector are subjected. Given that health inequalities remain a challenge in the DRC, it is imperative to introduce a policy of equity that encourages the commitment of health care providers. The United Nations Global Strategy for Women's, Adolescents' and Children's Survival sets out the objectives:
"Survive, thrive and transform", this strategy is comprehensive and equitable. (WHO 2019 P6).

That's why we need to ask ourselves some research questions.

I.2. Research questions

1. To what extent do the working conditions of healthcare providers in the Sankuru provincial health division affect patient satisfaction?
2. What are the current working conditions of healthcare providers in the Sankuru provincial health division?
3. To what extent are patients satisfied with the health facilities in Sankuru DPS? These questions form our research problem.

I.3. Hypothesis

The current working conditions of healthcare providers in the Sankuru DPS are said to be at the root of patients' disappointment.

I.4. Aim and objective of the study

The aim of our research is to contribute to improving the health of the community by increasing the effectiveness and efficiency of the healthcare system as part of the overall process of socio-economic development.

I.5. Specific objectives

- Determine the current working conditions of health care providers in the Sankuru provincial health division;
- Describe the level of satisfaction of patients who consult health facilities in the Sankuru DPS;

- Determining the impact of the working conditions of DPS healthcare providers

I.6. Background to study

Key words: Working conditions, healthcare provider, patient satisfaction, DPS Sankuru.

I.7. Definitions of key terms :

I.7.1. Working conditions: "generally speaking, working conditions refer to the environment in which employees live in their workplaces", Okongo (2014).

I.7.2. Healthcare provider: The word healthcare provider refers to the group of agents who practice the art of preventing, restoring or maintaining the health of an individual or a community. For example, doctors, nurses, dentists, laboratory technicians, physiotherapists, etc.

I.7.3. Division provinciale de la santé (DPS): The provincial health division is a decentralised administrative entity within the Ministry of Health. The Ministry of Health in the Democratic Republic of Congo is decentralised into 26 provincial health divisions, spread across the country's 26 provinces. Sankuru is one of the 26 provinces targeted by our study.

I.7.4. How important is the survey on this subject?

We believe that investigating this topic would provide us with thoughtful information on:

- The extent to which the working conditions of healthcare providers in the Sankuru DPS can have an impact on patient satisfaction;
- Current working conditions for healthcare providers ;
- The extent to which patients are satisfied, with a view to making responsible decisions that can improve ' well-being, in the overall context of reducing inequalities in care, with a view to achieving universal health coverage, as set out in the Global Compact.

CHAPTER II

REVIEW OF THE LITERATURE

This chapter of our work will deal with the literature relating to working conditions and the literature relating to patient satisfaction.

II.1. Working conditions :

The Sider computer programme explains working conditions as the environment in which employees find themselves in their workplaces. This brings to mind a number of important elements such as: remuneration, working hours, safety at work, social benefits, working climate, relations between colleagues, relations with superiors, work-life balance, opportunities for personal development, but also the arduousness of the work. An employee's health is negatively affected by poor conditions, whereas it is improved by favourable working conditions. As the saying goes, "a downcast spirit dries up the bones, but a cheerful spirit feeds the heart". This is why working conditions are governed by legislation and collective agreements.
(Sider, GPT-3.5. Turbo).

II.2. The relationship between work and health

Benjamin Franklin (2018) sees work initially as a painful constraint being perceived as a source of happiness, fulfilment and a basis that fosters relationships between humans.Etymologically, the word "labour" is derived from the Latin word "tripalium", meaning a three-legged instrument that was once used to inflict punishment on slaves. And in medicine, the pain of childbirth felt by mammals is also called labour, which is why hospitals have a labour ward. Benjamin Franklin (2018). The story of Adam and Eve in the Bible reminds us that Adam and Eve were punished after having offended God, for having transgressed divine law, and for having committed the original sin, that of having eaten the fruit of the forbidden tree, the tree of the knowledge of good and evil. God said to them: "By the sweat of your face you shall eat your bread" Gen3; 19. This idea of God punishing Adam and Eve does not mean that work is a punishment, but that it is the substance of work that is painful, and therefore the arduousness of work. This suggests that, in its place as an enriching, fulfilling activity that fosters relationships between human beings within a community, and enables people to prove their capacity for innovation and leave their mark the earth, work also has a negative side, namely its arduousness, suffering and burn-out syndromes. Benjamin Franklin's work showed that the link between work and health has been demonstrated since the 1700s, before the industrial era. (Benjamin Franklin 2018). Ronaldino RAMAZZINI, this professor of medicine and researcher at PADOU, discovered "Morbis artificum diatriba", the precursors of occupational pathologies, even before we could talk about the latter. There is no longer any doubt about the impact of work on people's health. Benjamin Franklin (2018). In the 19th , Louis René Villermé, a doctor from Paris, also published important documentation describing working conditions and their frightening effects on the health of exposed populations. In 1840, he published a table on physical and moral state of employees in cotton, wool and silk factories. Other 19th-century researchers also published shocking documents on "the cruelty and violence of assembly line work which destroys human health and dignity". Benjamin Franklin (2018). It should be noted that nowadays, studies examining the link between health, work and their repercussions on the customer are becoming rare, for reasons that require further study.

II.3. Literature review of the concept of work and health as perceived by other disciplines

For Benjamin Franklin, the notion of work and health is not perceived and interpreted in the same way by the different disciplines that take an interest in it. The sociologist sees it as a social relationship; the lawyer sees it as a relationship of subordination between employee employer, but this relationship must be governed by regulations or an employment contract; a set of tasks prescribed for performance, according to the ergonomist, (i.e. ergonomics looks at adapting work to man and not adapting man to work); while the economist sees it as a factor in the production of goods or services; and for the psychologist it is a confrontation with the outside world. Benjamin Franklin (2018). And in practical terms, work is paid professional activity with tasks to be carried out in an organisation, all governed by a contract of employment. Benjamin Franklin (2018).Health is a difficult concept to define, as it is not simply the absence of disease or infirmity.The World Health Organisation defines it as a state of complete physical, mental and social well-being and not merely the absence of disability or infirmity. (WHO, 1948).For legal scholar Hubert Seillans, "work is both the fruit and the source of the wealth necessary for health. Health itself is the source of wealth and quality work" (Hubert Seillans sd, Benjamin Franklin 2018). He goes on to say that health at work is simultaneously seen as a state, an objective, a means and an end.

II.4. Review of the literature on patient satisfaction

Excellent studies by JIHANE, S (2021) explain that customer satisfaction and customer experience are difficult to dissociate, especially when the two words are used together in a given context.For Pâlichon, 1999 and JIHAN 2021, "customer satisfaction is considered in marketing as a state of mind provided by an affective and cognitive evaluation process following a specific transaction". (Pâlichon, 1999; JIHANE, S. 2021, p 139; Okongo 2023 p 4/11).Customer satisfaction is a concept that is pragmatically understood as a one-dimensional sequence involving two extreme poles: the positive pole (very satisfied) and the negative pole (very dissatisfied). (Howard and Sheth, 1969; Oliver, 1980; JIHANE, 2021 p140).Measuring the customer experience is the same as measuring customer satisfaction.CAROLINE MERDIGER-RUMPLER's literature review (2009, p1-3) demonstrates how service elements contribute to inpatient satisfaction, using Llosa's "Tetraclass model (1997). This model appears to be the first to be tested in the field of human health, in short-stay surgery. This model was considered important insofar as it provided providers with new knowledge about the way in which patient satisfaction is structured, and also about decision-making on the choice of useful activities that could lead to improved patient satisfaction.In CAROLINE's view, two lines of thought have approached the literature on the concept of in-patient satisfaction, mainly: marketing literature focusing on consumer behaviour and service activities with a view to analysing the patient-consumer, and medical and hospital literature (again with a view to analysing the patient).

II.5. Literature on Customer Satisfaction in Marketing

Aurier & Evrard, Y. (1998), as cited by MERDINGER-RUMPLER, (2009). Point out in their work that authors have put forward several contradictory opinions and approaches to patient satisfaction in marketing, but they have agreed on the psychological state of the customer as the central and main characteristic of customer satisfaction, Aurier and Evrard, Y, (1998) ; MERDINGER-RUMPLER, (2009).

II.6. What does satisfaction mean as a psychological state?

It is considered a psychological state because it is the result of a process that integrates the cognitive and affective dimensions. Other authors have proposed the conative dimension of satisfaction as a third, although this is conceptually debatable (a psychological state that is totally different from its behavioural consequences, "the satisfaction of a phenomenon that is not directly observable". (Llosa, S. (1997)

II.7. Satisfaction as a Judgement of the Post-Consumer Experience

Even if the consumer's evaluation is based on the whole of their consumption experience, or if it only concerns part of the process, it can be either the purchase, the consumption or simply the use, the whole thing is called "post-purchase relative judgement". MEDINGER- RUMPLER, p2 (2009).In the case of service activities, the definition of satisfaction must take into account the experiential nature of the service and the need for the consumer to have experienced the service before making a satisfaction judgement. MEDINGER-RUMPLER, p2 (2009).

II.8. The literature on satisfaction as a relative assessment

The paradigm of the disconfirmation model provides an approach based on the process of satisfaction formation, which is founded on the following principles comparison of the consumer's subjective experience with comparison standards, also known as the initial reference base. Because of the limitations that have led to criticism of the expectations disconfirmation model, more than twenty comparison standards have been proposed. This led to the emergence of several competing explanatory theories. Today, it is accepted that these comparison standards produce different effects on satisfaction (these effects can be: linear or non-linear, positive or negative).While scientists find that the interest of this work is based on the evidence, from a managerial point of view its interest remains limited by the fact that the study focused on the disconfirmation of comparison standards, without taking into account certain important parameters, principally: role, importance and determinacy of the attributes making up the product or service offering. We have just recalled the important characteristics of satisfaction as defined in marketing. We want to focus on the satisfaction of a particular 'customer': the patient. MEDINGER-RUMPLER, p2 (2009).

II.8. Literature review of patient satisfaction as a comparative and multidimensional assessment

According to MEDINGER-RUMPLER, (2009), the abundant literature on patient satisfaction has proposed definitions which systematically integrate the "comparison" element of the training process without forgetting its multidimensional aspect which takes into account all the attributes which characterise the offer.

MEDINGER-RUMPLER, p2 (2009).

Generally speaking, patient satisfaction can be understood as the result a process evaluation and comparison of the service received by the various parties involved, and of the elements of the physical environment in which the healthcare service is delivered. Several studies, mainly Anglo-Saxon, have validated the parameters proposed by patients to evaluate their experience of hospitalisation. Despite the variety of evaluation dimensions proposed by researchers, four major groups of elements have been systematically retained. These are :

- "Interpersonal relations" (with different categories staff, medical, paramedical, nursing, administrative);
- "The physical environment (, rooms, restaurants, services).
- "The different procedures (reception, coordination of examinations, discharge).
- Technical quality of care

We should point out here that certain more technical elements are not taken into account in our literature review, in particular: clinical outcome, effectiveness of care, improvement, maintenance of health status, reduction in stress or pain, etc. In the light of all this literature, our study aims not only to take stock of current working conditions, but also the level patient satisfaction and the impact of working conditions on satisfaction, in order to provide a line of reference that will enable politicians to make adjustments when really need to improve the quality of patient care in the Sankuru DPS.It's difficult to talk about quality in general and the quality of care in particular (patient satisfaction), without recalling the notion of the ISO standard in healthcare organisations.

II. 9 The business standard in healthcare

In 2008, the Groupe entreprise en santé created this standard, which was endorsed by the Bureau de Normalisation Québécois, the CAN/BNQ 9700-800/2020 standard. The world's first ISO standard, it deals Prevention, promotion and organisational practices favourable to health in the workplace, commonly known as the Healthy Enterprise and therefore hospitals, clinics, health centres, etc. ...According to Lipari & Messier, 2020, as quoted by SAMUEL JULIEN in 2022, it is the very first standard in the world to deal with the promotion of overall health in the workplace (Lipari & Messier, 2020, SAMUEL JULIEN 2022 p19). After its latest revision in 2020, it is recognised in Canada as a National Standard and is one of the benchmarks throughout the country. Messier, 2020 confirms that in virons 40 health companies are already in possession of their health company certification. This is less than the number of ISO 9000-certified companies.What is the aim of this standard? The aim of this standard is to create working conditions conducive to the adoption and maintenance of good practices that are conducive to staff life and the sustainable improvement of employee health and well-being within the company. (Bureau de normalisation du Québec 2020; SAMUEL JULIEN 2022, p 20).The key word in this standard is prevention, and prevention is a priority. It must be put in place, promoted, maintained and improved organisational practices that are conducive to health. To achieve this, the standard requires :

- "Integrating the value of people's health into the company's management process.

• Creating or improving working conditions to prevent work-related illnesses and injuries;

• Creating conditions in the workplace that promote employee health and well-being;

• The implementation interventions that take into account both the needs of employees, which are identified through periodic data collection, and the company's challenges". (Bureau de normalisation du Québec, 2020; SAMUEL JULIEN, 2022). This standard enables stakeholders to create a climate of open, close collaboration for a healthier workplace. Its success is governed by three principles:

• "Shared responsibility for health between employees and workplace stakeholders;

•A firm, concrete and viable commitment from management;

•A close partnership between management, staff and all stakeholders". (Bureau de normalisation du Québec, 2020; SAMUEL JULIEN 2022, p 20/161). And if a healthcare wishes to obtain certification, there are three levels of certification:

Basic, Elite and Elite plus certification. These levels of certification differ in their requirements. The last two are more demanding than basic certification. The certification process for a healthcare company takes an average of 12 to 24 months. What sets it apart from other standards in other healthcare programmes is the importance attached to management commitment throughout the process. (SAMUEL JULIEN 2022).

CHAPTER III

METHODOLOGY AND METHOD

In every research project, it is compulsory to specify the methodology and data collection method(s) that will enable the research question to be answered. As far as our research subject is concerned, it seeks to answer the following research questions:

1. To what extent do the working conditions of DPS Sankuru employees affect patient satisfaction?

2. What are the current working conditions of healthcare providers in the Sankuru provincial health division?

3. To what extent are patients satisfied with the health structures in the Sankuru DPS? The positivist paradigm and the hypothetico-deductive, inductive and abductive approaches were chosen.

The philosophical approach (using the hypothetical-experiential category), based on ontological and epistemological hypotheses.

III.1. Definitions of concepts

III.1.1. Paradigm :

A video consulted on the blog METHODORECHERCHE.COM, which talked about the difference between a research methodology and a research method, explains paradigms as frameworks used by researchers as a basis for everything else they do. As explained by the blog, the contemporary meaning of the word is attributed to an American philosopher and physicist by the name of Thomas Kuhn. Thomas Kuhn sees a paradigm as a set of basic points of view and practices on which scientists agree at a given time.

For Guba and Lincoln (1994), as quoted in the above blog, paradigms are basic belief systems based on ontological, epistemological and methodological assumptions.

In other words, different types of research are based on different sets of beliefs; if we want to understand research, we need to examine the philosophy behind it. Just as the lenses of spectacles change the appearance of the environment, so a paradigm changes the direction of research. If you wear coloured glasses, everything becomes coloured, if you wear black glasses, everything becomes black, if you wear red glasses, the environment becomes red.

So lenses, or the paradigm we choose, change the way we see the world (Blog METHODORECHERCHE.COM).

III.1.2. Epistemology

Piaget, (1967), quoted by Daniel Le Garrec, (2023), sees epistemology as the study that seeks to construct valid knowledge halfway between objective idealism on the one hand and materialism on the other. To answer our research question, to what extent are the working conditions of employees in hospitals and health centres in the DPS Sankuru, we start from a set of ideas, valid and objective theories, towards the collection of information, then construct new knowledge. 13The study of philosophy that examines the nature, origin and limits of human knowledge is called epistemology. The word derives from two Greek words, episteme, meaning "knowledge", and logos, meaning "reason", hence the name "theory of knowledge". According to Cohen, (1996); as quoted by Daniel Le Garrec, (2023), epistemology is one of the branches of philosophy concerned with knowledge. Daniel Le Garrec, (2023).

III.1.3. Deductive approach

Gratton and Jon (2009), as quoted by Daniel Le Garrec, explain that the role of deductive approach is to test the explanation, test a theory, or test a predetermined hypothesis. Also known as deductive reasoning or deduction, this is the basic type of reasoning. This type of reasoning begins with a general statement or by a hypothesis, and then think about examining the possibilities of reaching a specifically thought-out conclusion. Systematic analysis of existing information will lead to acceptance or rejection of the hypothesis, with view to achieving the research objective, (Gill and Johnson, (2010) Daniel Le Garrec, p13 (2023).We believe that this approach will enable us explain the extent to which the current working conditions of employees in hospitals and health centres in the Sankuru DPS have an impact on patient satisfaction.

III.1.4. Inductive approach

According to Gratton & Jones (2009, cited by Daniel Le Garrec, 2023) p13, in contrast to deductive reasoning, the inductive approach requires the availability of the data collected to generate explanations, with the aim of exploring new theories. Flexibility is its advantage, in that the researcher is not obliged to follow predetermined theories every time.

III.1.5. Abductive approach

A third approach is called the abductive approach. In this approach, the process stems from a "surprising fact", also known as "puzzles", and the approach remains focused explaining the "surprising facts" or "enigmas". This is the "problem" that can arise when the researcher is confronted with an empirical phenomenon for which the existing theory has no explanation. Our study therefore intends to carry out this triangulation in order to achieve its objectives.

III.1.6. Data Collection Methods and Tools

To collect the data for our study on employee working conditions and patient satisfaction, we used a literature review with our checklist, which enabled us to explore a number of documents that provided us with the information we needed to write this dissertation.

III.1.7. Type of study

This is a secondary, analytical study focusing on employee working conditions and patient satisfaction. The aim of this study is to add to what is already known about the impact of working conditions on patient satisfaction.

III.1.8. Study site

As a secondary study, it is the result of a study published by POLITECHNIQUE MONTREALAISE, which is affiliated to the University of Montreal in Canada. The primary study used data from kept at "ALAYACARE", this Montreal-based company is their industrial partner, it has been responsible for collecting customer activity data since 2014, mainly data from home care centres in Canada, the United States and Australia.

III.1.9. Study population

The target group for this study would be employees of health facilities in the Sankuru provincial health division and patients temporarily residing there. However, following instruction from the university to conduct secondary research, our study population was switched to data extracted from the datasets of similar studies. The study population consisted of employees of 10 homecare agencies in Canada, the USA and Australia.

III.1.10. Sampling technique

Our study explored secondary data from a study which aimed to : Our study explored secondary data from a study that aimed to "qualify employee satisfaction within their jobs" (Guillaume 2020), in 10 homecare agencies that were targeted for experimentation. Namely: 6 were located in Canada, and had collected information from 72 customers, 2 located in the United States and had provided data from 24 customers and finally 2 in Australia, having 21 customers, in total 117 customers from 10 care agencies constituted the sample of the primary study source of our study. This data from different customers was taken from "ALAYACARE", the Montreal-based company responsible for collecting data on customer activities since 2014. And to meet the needs of our study, we had to re-examine these results from a comparative study across the different customers in the three countries.

III.1.11. Data collection and analysis tools

We used the Liker checklist to collect information that measured the impact of the employee's working conditions on patient satisfaction. We used SPSS mathematical software version 3.5.1 to analyse the data for our variables of interest.

Microsoft Word and PowerPoint were used to enter and present the study report.

III. What results have been achieved?

This study gave us 1/3 of the following results:

• The impact of working conditions on patient satisfaction was determined.

• The current working conditions of employees in health care facilities have not determined, and this requires an additional primary study to do so.

• The levels of satisfaction of patients who consult the medical care structures of the Sankuru DPS have not also been described and require a primary study geographically localising the subject, given that our variables of interest in relation to these results do not exist in the current databases of the country in general and the province in particular.

CHAPTER IV

PRESENTATION OF RESULTS

To find the results of our dissertation, which seeks to answer the question: to what extent can working conditions have an impact on patient satisfaction, we drew inspiration from Guillaume's work entitled:"From this dataset, we extracted the data for our six relevant variables. From this set of data, we extracted the data for our six variables deemed relevant. These data, which were again encoded and analysed using SPSS software, were processed as part of our study and were used to measure whether or not our research objectives had been achieved.

IV.1.LIST OF VARIABLES OF INTEREST

1. Remuneration

2. Working hours

3. Employee benefits

4. Relationships between department colleagues

5. Work-life balance

6. Opportunities for personal development

7. Patient satisfaction.

IV2.ANALYSIS MODEL USED

Statistical model of logistic regression

According to B. Karsh, B. Booske and F. Sainfort, (2005), as cited by Guillaume, simple logistic regression is the linear model used to determine the coefficients $b_o, ... b_n$ of the explanatory variables $x_1, ... x_n$ that best describe a variable y according to its formula.

The same author adds that: in the use of simple logistic regression, the variable y to be explained is binary. For our case, the objective is to determine the extent to which working conditions impact patient satisfaction. A. Y. Ng, (2004) added regularisation to the model by presenting his methods, using L1 and L2 norms. This latter method makes it possible to cancel out the coefficients of variables that provide little or no information in the prediction by adding a term to the likelihood function. In addition, the **scikitlearn** library offers a version that integrates the two models using the two standards, with an 11_ratio parameter that defines the importance of one over the other. This regularisation is called "Elasticnet" (Guillaume 2020 p 25).

Table I. Results of the logistic regression model

Term	Odds Ratio	95%	C.I.	Coefficient	S. E.	Z-Statistic	P-Value
Variable 1 remuneration	0,2926	0,2200	0,3893	-1,2288	0,1456	-8,4414	0,0000
Variable 2. Timetable of work	0,5766	0,4018	0,8273	-0,5507	0,1842	-2,9889	0,0028
Variable 3. Les social benefits	0,9305	0,8154	1,0618	-0,0720	0,0674	-1,0693	0,2849
Variable 4. Relationship between service colleagues	0,6109	0,4591	0,8129	-0,4928	0,1458	-3,3811	0,0007
Variable 5. Work-life balance privacy	1,6200	1,1462	2,2806	0,4824	0,1765	2,7332	0,0063
Variable 6. Development opportunities staff	0,6681	0,5096	0,8759	-0,4033	0,1382	-2,9194	0,0035
CONSTANT	*	*	*	1,7483	0,1694	10,3179	0,0000

Comments: The above table shows that the relationship between working conditions and patient satisfaction is statistically highly significant at the $\alpha \leq o, o5\%$ threshold, so social and economic characteristics have a positive influence on patient satisfaction, since the p value is well below 0.05. The odds ratio is 0.2926. The odds ratio is 0.2926.

Working hours and work-life balance were statistically significant, with p-values $\leq 0.05\%$ and odds ratios of 0.5766 and 1.6200 respectively, while fringe benefits, relations between colleagues and

29

opportunities for personal development were also statistically highly significant. These variables have influence on patient satisfaction, as their p-values were found to be largely$\leq 0.05\%$, with odds ratios of : 0.9305; 0.6109 and 0.6681.

CHAPTER V

DISCUSSION, CONTRIBUTIONS OF THE RESEARCH, LIMITATIONS OF THE RESEARCH AND AVENUES FOR FURTHER RESEARCH

V.1. Theoretical and managerial input

From a theoretical point of view, this statistical logistic regression model us with information confirmed or validated several hypotheses of the other mathematical models and theories previously explored in relation to working conditions and patient satisfaction. These results confirm, among other things, JAMAL and WARIT's assertion (18 January 2024) that "a satisfied employee equals a satisfied customer". A satisfied employee becomes more active and productive. The company has a major responsibility in managing employee satisfaction in order to maximise results, improve the quality of the community's health, and improve the efficiency and effectiveness of the healthcare system.

From a managerial point of view, the results of this research provide sufficient evidence of the influence of working conditions on patient satisfaction, and confirm the results of the work of JAMAL and WARIT, (18 January 2024), who argue that the impact of corporate social responsibility (CSR) on the customer must be achieved first and foremost through employee satisfaction. These studies show that Moroccan citizens are right to be concerned about companies that mismanage their employees, because of the poor quality products they produce and put on the market for the benefit of the population. Moroccan citizens have sufficiently understood the link between a

31

quality product and employee satisfaction. A satisfied employee will devote his or her soul and conscience to performing work that produces quality results, whereas a dissatisfied employee runs the risk of deviating from the supply curve or indulging in supply-induced demand practices. It would therefore be appropriate for companies that have aligned themselves with the logic of implementing universal health cover to take into account the need to protect the health of their employees. employee satisfaction management, if and only if companies want to improve the quality of care offered to patients. Work should not just be seen as a painful constraint, but also as a source of enrichment and fulfilment for the employee and society as a whole.

The results of this research should be of interest to policies on the improvement of working conditions whenever quality improvement is discussed, given quality improvement implies customer satisfaction, and this satisfaction depends mainly on employee satisfaction.

The influence of working conditions on patient satisfaction demonstrated by this research leads us to believe that negligence, lack of commitment, ill-treatment inflicted on patients, the imbalance between supply of care and that of demand, care demand practices induced by carers, as well as all other practices deviating from the medical care supply curve, observed in health care facilities in the Democratic Republic of Congo in general and in the DPS of Sankuru in particular, are the result of dissatisfaction among employees in this sector, and that the solution lies in their satisfaction.

V.2 Limits of the research and avenues of research

From a theoretical point of view, our research merits additional research geographically locating the province of Sankuru in the Democratic Republic of Congo, mainly to collect primary data that can answer the questions not answered by this research, in particular the following questions: 1. What are the current working conditions of health care providers in the Sankuru provincial health division? 2. To what extent are patients satisfied at Sankuru DPS health facilities? These questions remain unanswered as our research was switched to a secondary search and found no information geographically localising the subject to the province of Sankuru. The subject does not yet seem to be covered in the province. From a methodological point of view, it would seem appropriate to replicate and experiment with this subject locallyusing the Likert scale to capture geographically localised realities and look for ways of resolving them, given that most of our African governments, principally the Ministry of Health, Hygiene and Prevention in the Democratic Republic of Congo, has not yet incorporated variables measuring employee and patient satisfaction into its database. While the country is still fighting disease by developing thousands of disease control indicators, the variables measuring employee satisfaction seem to have been orphaned or simply obsolete, when they should have been put together to better assess and understand the level of improvement in community health.

CONCLUSION

Overall, the aim of this study is to contribute improving the health of the community by increasing the effectiveness and efficiency of the health system as part of the overall process of socio-economic development. Its objectives were to determine the impact of the working conditions of healthcare providers in general, and those of the Sankuru provincial health division in particular, on patient satisfaction. To achieve this, we started from a statement about the health problem in sub-Saharan Africa, which is precarious and of which our country, the Democratic Republic of Congo, is a part, as mentioned in the World Health Organisation's report on the technical consultations and country action plans held in Adi-Abeba from 22-26 July 2019. Based on these observations of the failings of our healthcare system in relation to the dissatisfaction of patients and employees in the public sector, our hypothesis predicted that the current working conditions of healthcare providers in the Democratic Republic of Congo in general and the Sankuru DPS in particular would be at the root of patient dissatisfaction. After statistical test analysis, mainly logistic regression of the secondary data collected revealed that the relationship between conditions and patient satisfaction is statistically highly significant at the α threshold. $\leq o, o5\%$ thus the social economic characteristics mainly the remuneration of the employees positively influence the satisfaction of the patients, because, the p value being largely lower than 0,05% and the ratio of coast is 0,2926. From this, we can deduce that in sub-Saharan Africa in general and the DRC in particular, the working conditions of health sector employees are at the root of patient dissatisfaction, and employee dissatisfaction has a negative impact on patient satisfaction. Our research has shown that our healthcare system has developed more indicators related to disease control, hygiene, prevention, etc. Unfortunately, our exploration has noticed that no employee satisfaction indicators are available in our databases! This indicates a lack of interest in, or knowledge of, the value of indicators for monitoring employee satisfaction, which are key indicators for assessing and improving the quality of health care for a population, with a view to achieving universal health coverage and complying with the CAN/BNQ 9700-800/2020 and ISO 9000 standards. We will not be able to successfully implement free healthcare without taking into account the satisfaction of healthcare staff. The results of our study confirmed an adage from Sankuru says: "Otemanyanga hadjolemu nkamba", which translates into French as: "A heart broken by worry doesn't get the job done". An employee's dissatisfaction breaks his or her heart, and despite the hardship it seems to be unappreciated by the employer and therefore obsolete. The question of the working conditions of public health sector employees and its impact on patient satisfaction is an intriguing one that is worth examining in further research. Our study experienced limitations that did not allow us to measure not only the current conditions of Dps sankuru employees but also the current satisfaction levels of patients who consult these care facilities, due to the lack of such data in provincial databases. Because of this, our study would be important be conducted empirically to arrive at local evidence or to be limited to secondary data analysis where most variables of local interest not available. We would recommend that the governance and leadership of our country's healthcare system should incorporate indicators for periodic monitoring of employee satisfaction as well as periodic patient satisfaction surveys. This periodic comparison will boost the improvement in the quality of care which, in turn, will increase the effectiveness and efficiency of the healthcare system as part of the overall process of socio-economic development.

REFERENCES

1. A. Y. Ng,"Feature selection, l1 vs. l2 regularization, and rotational invariance," in Proceedings of the Twenty-First International Conference on Machine Learning, ser. ICML '04. New York, NY, USA: Association for ComputingMachinery, 2004, p. 78.

2. Benjamin Franklin et al (2018)   Understanding the relationship between work and health: friend or foe? La Santé et le Travail (2nd edition) : 10 étapes Pour une Prévention Efficace Dans L'entreprise, Arnaud Franel Editions, 2018. ProQuest Ebook Central, http://ebookcentral.proquest.com/lib/UNICAF/detail.actio n?docID=5651343. Created from UNICAF on 2024-05-12 13:56:55.

3. Daniel Le Garrec. (2023) Introduction to Scientific Research, unicaf university.

4.GUILLAUME VERGNOLLE. (2021). Tool for diagnosing and preventing employee dissatisfactionUniversité de Montréal, Canada, Distributed by ProQuest LLC

5. SEBAI. J. (2021), Kroger et al (2007), Donabedian (1980). From experience à to satisfaction patients : Force improvement in France, Université Paris-Saclay, UVSQ, Larequoi Versailles, France.

6. MERDINGERRUM PLER. (2009) Contribution of service elements to inpatient satisfaction: an application of the tetraclass model, marketing decision no. 53 January-March 2009-43.

7.MUSUNGAYI (8Feb2024). Suspension of the medical directors of two hospitals in Kinshasa, DRC, Congolese Press Agency/ACP POLITICO.CD

8.Organisation World Health Organization. (2019). Technical

consultation and country action plan, from 22- 26July 2019 in ADIS ABEBA. P1

9. Okongo (2023), Plichon. (1999), JIHANE, S. (2021). Impact of Health Professionals' Working Conditions on Patient Satisfaction Case Study of the Provincial Health Division (Dps) of Sankuru, Republic of Congo

Démocratique Du Congo/RDC, Université unicaf.

10. Youssef JAMAL, Rajaa FAIK, and Mohamed WARIT. (2024). Littérature review on the existing link between human resource Management and social performance, Université Hassan II - Casablanca- MAROC ISSN : 2658-8455 Volume 5, Issue 2 (2024), pp. 149- 160. © Authors : CC BY-NC-ND

APPENDICES

Interview guide on the working conditions of healthcare providers and patient satisfaction.

*Hello, my name is **I'm** a researcher in the field of improving the health and well-being of everyone. Thank you for your participation. The purpose of today's interview is to find out more about the working conditions of health facility providers in the Sankuru DPS and the impact these have on patient satisfaction. We would like to gain a deeper understanding of the working conditions in your environment, exploring the views, challenges and factors that can facilitate the integration of the occupational health programme into the local health system. This interview will cover a variety of questions relating to working conditions for providers and a second part relating to patient satisfaction. You are asked to respond according to your level of conviction:(Strongly agree, Agree, Somewhat agree, Neither agree nor disagree, Somewhat disagree, Strongly , Don't know, No answer). Patients, on the other hand, will respond according to a satisfaction scale: (Very satisfied, satisfied, not at all satisfied, very dissatisfied, etc.). I'd like to record this interview; would that be OK with you (turn on the recorder and repeat this question, for the recording if possible).*

Before we start, do you have any questions for me?

First of all, could you please tell me a bit about yourself?

Explorer : name province

Health zone

Health area

Position

Seniority

Date 2024

The Investigator, First name, surname and signature: Tel :

SECTION I. QUESTIONS FOR ASSESSING CURRENT WORKING CONDITIONS

To what extent do you agree with the following statements?								
	I totally agree	Agreed. Somewhat	Neither agree or disagree	Somewhat disagree	Not at all Agreed	I don't know	No answer
My work is in line with my personal life (no difficulty in reconciling my life and my work). work).								
I am not stressed at work								
My work environment is conducive to my safety and protects my mental health.								
I am not intimidated in my workplace. There is no malicious, offensive, intimidating or insulting behaviour in my environment. of work.								
My work environment guarantees my health physical.								
My work has no role conflict								
My has always been evaluated and adjusted to suit me. (ergonomics).								

My job guarantees me great prospects promotional								
My work is coordinated with that of others group members								
Our combined work is harmonised								
The conditions of my work drive me to commit myself to the well-being of patients and his family.								
As part of my working group, I am committed to making my contribution for the good of the Group. patient								
Mutual trust reigns in our Group of work								
I can take part in sporting activities at my place of residence. work								
Do you have anything else to add to improve or maintain working conditions?Answer(s) :								

SECTION II. QUESTIONS FOR ASSESSING PATIENT SATISFACTION.

	How satisfied are you with the hospital environment?							
		Very satisfied	Satisfied	Somewhat satisfied	Neither satisfied nor dissatisfied	Somewhat dissatisfied	Not at all	I don't know	No answer
Patient reception	Accessibility of training sanitary								
	Welcome by service providers (administrative and techniques).								
	Welcome booklet clear								
Patient care (medical and paramedical).	Clarity of information and of responses received at questions.								
	Sharing the decision medical.								
	Support for professionals, attentive listening and								

	respect								
	Help given suffering increases or urgency.								
	Respect for privacy								
	Secret undisclosed professional								
	Care in cases of pain								
	Support for others discomfort								
Accommodation and meals	Comfort of the room								
	Quality and quantity of meals.								
Discharge and its organisation	Explanations received regarding the resumption of activities, signs or complications that could lead to the need to contact new the								
	training health.								
	Information relating to post-discharge monitoring, the appointment.								
	The cost of healthcare (medical and nursing) does not fall on me or my family. family.								
Is there anything else you would like to add to improve or maintain the quality of care?Answer(s) :									

Printed by Books on Demand GmbH, Norderstedt / Germany